ng

THE **TESTING** SERIES

FIREFIGHTER **INTERVIEW QUESTIONS & ANSWERS**

THE **TESTING** SERIES
expert advice on test p

Orders: Please contact How2become Ltd, Suite 2, 50 Churchill Square Business Centre, Kings Hill, Kent ME19 4YU.

Telephone: (44) 0845 643 1299 - Lines are open Monday to Friday 9am until 5pm. You can also order via the email address info@how2become.co.uk.

ISBN: 9781907558405

First published 2011

Typeset for How2become Ltd by Molly Hill, Canada.

Printed in Great Britain for How2become Ltd by Bell & Bain Ltd, 303 Burnfield Road, Thornliebank, Glasgow G46 7UQ.

CONTENTS

PREFACE
BY AUTHOR RICHARD MCMUNN

Over the past few years I have coached and helped literally hundreds of people to prepare for different selection processes and interviews. My particular niche, however, is preparing people for a career in the Fire Service. I spent 17 years in Kent Fire and Rescue Service, during which time I worked at the Fire Service Training Centre. During my stint at the Training Centre I was involved in:

- Running firefighter trainee courses (4 in total)
- Teaching serving firefighters how to wear breathing apparatus
- Teaching serving firefighters new and amended technical operational procedures
- Marking application forms for new applicants
- Sitting on interview panels assessing potential trainee firefighters to join the job

I have no doubt that all of this experience has helped me to become successful as an author, coach and mentor. I have a genuine passion for helping people like you to achieve their goals. However, none of this is possible without the co-operation of the student; in this case, you. It is crucial that you spend plenty of time on your firefighter interview preparation. You must leave no stone unturned and prepare your responses for every question in this book. Follow my advice carefully and I guarantee you will be fully prepared and confident when you walk through the interview room door on the day of your interview.

I joined the Fire Service on January the 25th 1993 after completing four years in the Fleet Air Arm branch of the Royal Navy. In the build up to joining the Fire Service I embarked on a comprehensive training programme that would see me pass the selection process with relative ease. The reason why I passed the selection process with ease was solely due to the preparation and hard work that I had put in during the build-up.

I have always been a great believer in preparation. Preparation was my key to success and it is also yours. Without the right level of preparation you will be setting out on the route to failure. The Fire Service is very hard to join, but if you follow the steps that I have compiled within this guide then you will increase your chances of success dramatically.

Remember, you are learning how to be a successful candidate, not a successful firefighter!

The Fire Service has changed a great deal over the past few years and even more so in how it assesses potential candidates for firefighter positions. When I joined in 1993 it helped if you were 6ft tall, built like a mountain and from a military background. Things have certainly changed since then and rightly so. Yes, the Fire Service still needs people of that calibre but it also needs people who represent the community in which it serves. It needs people from different backgrounds, different cultures, different ages, different sexual orientations and different genders. Basically, the community in which we live is diverse in nature and therefore so should the Fire Service be if it is to provide a high level of service the public deserve.

Most of us will thankfully go through life never having to call upon the Fire Service. Those who do call on the Fire Service expect their firefighters to be physically fit, professional and highly competent in their role.

During my time in the Fire Service I attended hundreds of different incidents ranging from property fires, road traffic collisions, chemical incidents, ship fires and even rail accidents. During every single one of them I gave my all, and so did my colleagues. During your time in the Fire Service you will experience many highs and many lows. The highs will come from your ability and influence to save a person's life and naturally the lows will come from the people whom you sadly could not help. How you handle the low points of your career is crucial. Fortunately, you will experience an amazing level of comradeship during your career that is extremely rare and is not normally found in other jobs or professions. It is this high level of comradeship that will get you through the low points.

The men and women of the UK Fire Service carry out an amazing job. They are there to protect the community in which they serve and they do that job with great pride, passion and very high levels of professionalism and commitment. They are to be congratulated for the service they provide.

As you progress through this guide you will notice that the qualities required to be a firefighter are a common theme. You must learn these qualities and also be able to demonstrate them at interview if you are to have any chance of successfully passing the selection process.

I wish you all the very best in your pursuit to passing the firefigter interview. Now, let's get started!

Best wishes,

Richard McMunn

Richard McMunn

P.S. I am now running a series of 1 day training courses to help people like you prepare for the firefighter selection process. Visit my website below to find out more:

FIREFIGHTERCOURSE.CO.UK

CHAPTER 1
ABOUT THE FIREFIGHTER INTERVIEW

During this section of the guide I want to answer some of the more common questions people ask me. The answers will help you to hone your preparation in the right areas.

Each stage of the application process is very important but you will probably find that this is the one stage that causes you the most nerves. In fact if you don't feel nervous leading up to and during your interview, then you are less likely to perform to your maximum capability.

If you have reached this stage of the selection process then the Fire Service are interested in employing you and they want to meet you face to face in order to see what you are like as a person, and also whether you really do have the skills required to become a firefighter.

Before I provide my sample interview questions and tips, let me answer a question that I get asked time and again by firefighter applicants: "What's the best way to prepare for the firefighter interview?"

Here's my response:

"The firefighter interview is used as a means to assess your potential to become a firefighter. Fitness is very important to the role but it is not the 'be all and end all'. Yes you must demonstrate a good level of fitness, but there are more important elements to demonstrate.

The firefighter interviewers will be looking for 'evidence' of where you can match the assessable qualities. The word evidence is pivotal and I suggest you have it at the forefront of your mind during your preparation. You have to provide as much evidence as possible as to how you match the assessable qualities. The reason for this is simple: if you can provide evidence then there is a far greater chance of you succeeding as a firefighter. Anyone can say that they are a good teamworker, or good at working with people from different backgrounds; however, providing evidence to back it up is a different thing altogether. So, during your preparation you should concentrate on providing specific examples of where you can meet the qualities being assessed, more on this later.

You should also think very carefully about why you want to become a firefighter. It's all well and good saying that you've wanted to do this since you were a little boy or girl, but responses like this will gain few marks. You have to have a genuine reason for wanting to join; something like 'wanting to make a difference to your community' is a much better reason for wanting to join. Firefighters are caring people, which is why they have so much respect amongst members of society.

Finally, and this is just a big a factor than any other, your 'likability' will be key to your success. Yes, there are rules and guidelines that the interviewer must adhere to when interviewing potential candidates, but you can never take away the personal element. You should try hard to come across as a genuine, reliable, professional and a conscientious individual. Do not be arrogant or over-confident at the interview and always try to demonstrate your willingness to learn and be part of the wider Fire Service team."

If I was preparing for the firefighter interview today, I would first of all ask myself the following three questions:

Q1. What areas will the Fire Service assess me on during the interview?

Q2. What would they expect to see from successful interview candidates?

Q3. Can I provide 'evidence' of where I meet the assessable qualities?

I will then write down my perceived answers to these questions and I get the following responses:

A1. They will probably ask me questions that are based around the firefighter Personal Qualities and Attributes. They will require me to provide 'specific' examples of where I already have experience in each of the PQA areas. They will also want to know why I want to become a firefighter and in particular why I want to join their Fire Service. They may also want to know how I have prepared for the selection process and what qualities I believe I have that will be relevant to the role.

A2. They would expect successful candidates to provide specific examples that relate to the PQA related questions. They would also expect my responses to the interview questions to be concise, relevant, well-structured and in a logical sequence. They would also want to see that I have put a large amount of work into my application and that I am genuinely motivated and committed to becoming a firefighter within their organisation. They will want to see that I have gone out of my way to find out about the role of a firefighter, including the proactive side of the role.

A3. I would need to think seriously about the type of evidence I have to demonstrate that I would make a good firefighter. For example, one of the assessable qualities required to become a firefighter is that of working with other people; or in other words, teamwork. I would need to be able to provide a number of specific examples of where I have worked as part of a team and demonstrated my competence in this area. I would also need to write down specific examples of where I can match the other assessable PQA areas. This will mean getting a copy of the Personal Qualities and Attributes that are relative to the role and working through them carefully, thinking of examples of where I match each of them.

Now that I have my answers I will set out another simple plan that dictates exactly what I am going to do put those answers right. In this particular case it will look something like this:

- I will learn the PQAs thoroughly and I will try to think of specific examples of where can match each and every one of them. I will use the STAR method when constructing my responses to the interview questions which will allow me to put them across in a concise and logical manner. I will think of genuine reasons why I want to become a firefighter and I will make sure that I can provide good reasons as to why I want to join their particular Fire

Service. I will also write down exactly what I have done to prepare for the selection process and also the qualities I believe I possess that would be suited to the role.

- I will visit my local fire station and speak to the firefighters about their role. I will also visit the website of the Fire Service I am applying to join and learn about the proactive side to the job in addition to the reactive side.

During the interview you will need to demonstrate that you have the potential to become a firefighter and this is what the panel will be looking for. You will have learnt a tremendous amount about the Fire Service and what the role of the firefighter actually involves in the build up to the interview.

Your preparation should start weeks before your interview date, not the night before!

FREQUENTLY ASKED QUESTIONS

The interview is designed to assess your knowledge of the firefighter's role and in particular how you can meet the Personal Qualities and Attributes that are relevant to the role. Whilst the majority of questions are usually based around the PQAs, you should also prepare for the more generic type of interview questions that are contained within this guide.

TIP: Read and understand the Firefighter Personal Qualities and Attributes before you attend the interview and be ready to provide specific examples of where your skills meet each one.

How long will the interview last for?

The interview should last no longer than 1 hour. Usually between 45 and 50 minutes is the norm but this will depend on the length of your responses. If you give short answers then it could be over in 10 minutes. My advice is to respond to the questions sufficiently to provide enough quality evidence. Don't waffle, but instead provide good, solid responses. The panel will ask you to stop once they have heard enough.

What should I wear to the interview?

You will not normally be assessed on your dress but I strongly advise that you wear a formal outfit such as a suit. Make sure your shoes are clean and polished and do not wear white socks or ones with cartoon characters on them! Remember that you are applying to join a disciplined service. If you cannot dress smart then how will you be able to wear your uniform with pride? Firefighters are role models for the service they work for. You should have good personal hygiene and be capable of wearing your uniform in a smart and presentable manner.

What is the purpose of the interview?

The main purpose of the interview is to talk about you and your interest in becoming a firefighter, and then to look at some areas of your experience in more detail. Questions will be based around the assessable qualities, as well as the more common types of question, such as why you want to become a firefighter and what you have to offer.

Is everyone asked the same set of questions?

Yes they are. This is so that the process is fair and everyone gets the same chance. Therefore, you should refrain from telling anyone the questions you were asked at the interview as they could be your direct competition!

How many people will be on the panel?

This depends on the service you are applying to join. There could be 2, 3 or even 4 people on the panel. They could be a mix of uniformed personnel and non-uniformed personnel. There will usually be someone on the panel from the Human Resources department to ensure the interview is fair and consistent in addition uniformed members of staff. When responding to the interview questions you should look at every member of the panel and not just the person asking the question(s); this is good interview technique.

There will normally be one member of the panel writing down your responses to the questions. Don't be put off by this. This simply allows the panel to discuss your performance at the end of the interview.

How will the interview commence?

The panel will ask you to sit down and they will introduce themselves to you. There will be a glass of water on the table in front of you and you will be asked to make yourself comfortable. Whilst it is ok to make yourself feel comfortable, do remember that you are being assessed so make sure you watch your interview technique.

The panel will then explain the purpose of the interview to you. They will state that they are going to ask you for specific examples of what you have done in different situations. It is okay to draw from examples from home, work, school, college or hobbies. It is advisable that you draw from a variety of different experiences.

How will I be assessed?

You will normally be assessed against the Personal Qualities and

Attributes that are relevant to the role of a firefighter but this will very much depend on the Fire Service that you are applying to join.

The key assessment areas might be as follows:

- Commitment to excellence
- Commitment to development
- Commitment to diversity and integrity (this area carries the most importance)
- Communicating effectively

The Fire Service will normally write to you before the interview and inform you of the assessable areas. My advice is to prepare responses for every single assessable quality. This will ensure that you are fully prepared for every eventuality. Remember, evidence equals points!

TOP TIP

Do not use jargon. Make sure you communicate in language the panel can understand.

Before you get into the formal assessment questions, the interview panel will more often than not ask you a set of 'warm-up' questions. These will last approximately 10 minutes and it is your chance to get off on the right foot.

What warm-up questions will I be asked?

The warm up questions will vary; however, here are a few examples of ones that have been used previously.

Q. How was your journey here today?

Try to answer this question more than just 'okay' or 'good'. Use the warm-up questions as a chance to communicate with the panel. For example:

"The journey was very good thank you. Before I came for the interview today I planned my route and checked on the news to see if there were any traffic problems or hold ups that could have delayed my travel time. I managed to avoid a tailback on the motorway, so thankfully the journey was very relaxing. I understand that being a firefighter involves an ability to plan and prepare and I believe I am very good at doing this."

Q. What interests you about becoming a firefighter?

This is a great opportunity to get the interview going along the right lines! The Fire Service offers teamwork, opportunities to help and support the community and the chance to work in a diverse workforce etc. Be positive in your response and think carefully about the role of a firefighter.

"There are many aspects of the role and the service that interest me. To begin with I am attracted to caring nature of the role. I enjoy helping and educating people and I would like to make a difference to the community. I have some experience of working in the community and I thoroughly enjoyed it. I am also attracted to the role because there is a chance to make a difference, both within the service and within the community. I believe I can contribute to the service by being professional, hardworking and conscientious and I also believe I can contribute to the community due to my patience, understanding nature and professionalism. In the whole, I am attracted to the job because it is a career that will give me the opportunity to achieve my full potential and by doing that others in the community will benefit."

Q. How did you hear about the fact that we are recruiting?

Try to show here that you are keen and have been keeping in touch with your local fire station and the service's website to check for recruitment updates, keeping your eye on the local paper etc.

"I have always kept a keen eye on the Fire Service to see when you are recruiting. This has involved keeping weekly checks on the Fire Service website and looking for vacancies in my local newspaper. I have visited my local fire station on two occasions and discussed recruitment opportunities with the local firefighters. I have been considering applying for the role for a number of years now and have always maintained an interest in the service."

Q. What steps have you taken to find out more about the role of a firefighter?

Again, maybe you have visited the local fire station, visited their website, attended a course, bought and studied a book or DVD etc.

"I have been working hard for the last 6 months when I heard that you might be recruiting sometime soon. I started off by visiting my local fire station to see what changes there had been to the role. This allowed me to gain a thorough understanding of how the firefighters work to improve the local community. I then obtained a book about the firefighters role to learn some more. I have been studying the Fire Services website to find out about what you have been doing in the local community and the type of campaigns you have been running. In particular, I noticed that you have been promoting barbeque fire safety messages as it is fast approaching the summer season. Finally, I purchased three psychometric testing books as I wanted to improve my ability to work with numbers. This enabled me to improve by ability in preparation for the written tests."

Q. Can you tell me about some of your interests or hobbies?

Try to tell the panel something about you that is interesting. Maybe you play sports or outdoor activities. Even if you just enjoy walking to keep fit, this is still positive.

"I am a keen keep fit enthusiast and spend 4 evenings a week working on my fitness and stamina levels. I particularly enjoy the indoor rowing machine and currently hold the record at my gym over 2000 metres. I also enjoy carrying out charity work and recently ran the London Marathon for a local children's charity. Finally, I play the guitar to level grade 5. I have worked hard over the last few years to achieve the grade by working hard to improve myself."

You could be asked a whole host of warm-up questions. Just make sure you provide positive answers and use the questions as an opportunity to get into 'interview mode'.

Once the warm-up questions are complete, the interviewer may take a drink of water at this point. You may also take a drink of water and compose yourself, ready for the formal interview questions. You will then be informed that you are moving into the formal part of the interview. A member of the panel will say something along the lines of 'OK, we are now moving on to the first PQA, which will last approximately 10 minutes'.

Please note – the PQA questions can come in any order.

THE MOST IMPORTANT PIECE OF ADVICE I CAN GIVE YOU

Throughout this guide so far I have made reference to the assessable firefighter qualities and attributes. During the interview there is a strong possibility that you could be asked questions relating to these.

If I were preparing for the firefighter interview right now I would take each PQA individually and prepare a detailed response setting out where I meet the requirements of it. Your response to each question that relates to the PQAs must be 'specific' in nature. This means that you must provide an example of where you have already demonstrated the skills that are required under the PQA. Do not fall into the trap of providing a 'generic' response that details what you 'would do' if the situation arose.

Try to structure your responses in a logical and concise manner. The way to achieve this is to use the 'STAR' method of interview question response construction:

SITUATION

Start off your response to the interview question by explaining what the 'situation' was and who was involved.

TASK

Once you have detailed the situation, explain what the 'task' was, or what needed to be done.

ACTION

Now explain what 'action' you took, and what action others took. Also explain why you took this particular course of action.

RESULT

Explain what the outcome or result was following your actions and those of others. Try to demonstrate in your response that the result was positive because of the action you took.

Finally, explain to the panel what you would do differently if the same situation arose again. It is good to be reflective at the end of your responses. This demonstrates a level of maturity and it will also show the panel that you are willing to learn from every experience.

THE DIFFERENT TYPES OF INTERVIEW QUESTIONS

Basically there are two different types of interview questions that you could be asked. I will try to explain each of them and what they mean:

1. Generic questions about you and your knowledge of the Fire Service and the firefighter's role.

Generic questions can be in any format. There is no particular structure to this type of question but they are generally far easier to respond to. Examples of generic questions would include:

- Why do you want to become a firefighter?
- What has attracted you to this Fire Service in particular?
- What have you learnt about the role of a firefighter?
- Why should we choose you against the other applicants?

Generic questions are becoming less and less popular during the firefighter interview and the majority of Fire Services are now using 'PQA' based questions. However, it is still important to prepare for the generic type of questions.

2. PQA questions.

This type of question is becoming more and more popular during the firefighter interview and you should certainly concentrate the majority of your preparation in this area. The personal qualities and attributes are generally the 'blueprint' for the role of a firefighter. The interview panel will want to know whether you already have these skills and that you can give examples to demonstrate your use of them in your responses to the questions.

CHAPTER 2
SAMPLE INTERVIEW QUESTIONS & RESPONSES

Within this section of the guide I will provide you with a large number of sample interview questions that have been used during firefighter selection interviews. Having asked some of these questions myself to potential candidates, I will also provide brief details on what I believe makes a strong response, and what I believe makes a weak one.

Whilst you will not be asked every question that follows, they will all give you a great hand during your preparation and I strongly recommend you prepare a response for each and every one of them. In order to assist you in your preparation I have provided a blank space following many of the questions for you to write down your sample response.

Plenty of hard work and determination is needed here, so be prepared to knuckle down and put in the effort.

QUESTIONS BASED AROUND THE PERSONAL QUALITIES

The first PQA deals with working with others and will show the extent to which you work effectively with other people in the community.

Q1. *Tell me about a time when you have contributed to the effective working of a team?*

How to structure your response:

- What was the size and purpose of the team?
- Who else was in the team?
- What was YOUR role in the team? (Explain your exact role)
- What did you personally do to help make the team effective?
- What was the result?

Strong response

To make your response strong you need to provide specific details of where you have worked with others effectively, and more importantly where YOU have contributed to the team.

Try to think of an example where there was a problem within a team and where you volunteered to make the team work more efficiently. It is better to say that you identified there was problem within the team rather than that you were asked to do something by your manager or supervisor.

Make your response concise and logical.

Weak response

Those candidates who fail to provide a specific example will provide weak answers. Do not fall into the trap of saying 'what you would do' if this type of situation arose.

Sample response

"I like to keep fit and healthy and as part of this aim I play football for a local Sunday team. We had worked very hard to get to the cup final and we were faced with playing a very good opposition team who had recently won the league title. The team consisted of 11 players who regularly spend time together during training sessions and at social events. After only ten minutes of play, one of our players was sent off and we conceded a penalty as a result. Being one goal down and 80 minutes left to play we were faced

with a mountain to climb. However, we all remembered our training and worked very hard in order to prevent any more goals being scored. Due to playing with ten players, I had to switch positions and play as a defender, something that I am not used to. Apart from being a defender I felt my role was to encourage the other players to keep going and to not give up until the final whistle had sounded. All the other players supported each other tremendously and the support of the crowd really pushed us on. The team worked brilliantly to hold off any further opposing goals and after 60 minutes we managed to get an equaliser. The game went to penalties in the end and we managed to win the cup. I believe I am an excellent team player and can always be relied upon to work as an effective team member at all times. I understand that being an effective team member is very important if the Fire Service is to provide a high level of service to the community in which it serves. However, above all of this, effective teamwork is essential in order to maintain the high safety standards that are set."

Now take the time to use the blank space on the following page to prepare your own response to this question.

QUESTION 1

Tell me about a time when you have contributed to the effective working of a team?

Q2. *Tell me about a time when you helped someone who was distressed or in need of support?*

How to structure your response:

- What was the situation?
- Why did you provide the help? (Whether you were approached or you volunteered – TIP: It is better to say you volunteered!)
- What did you do to support the individual?
- What specifically did you do or say?
- What was the result?

Strong response

Again, make sure you provide a specific example of where you have helped someone who was in distress or who needed your support. Try to provide an example where the outcome was a positive one as a result of your actions. If the situation was one that involved potentially dangerous surroundings (such as a car accident), did you consider the safety aspect and did you carry out a risk assessment of the scene?

Weak response

Candidates who provide a weak response will be generic in their answering. The outcome of the situation will generally not be a positive one.

Sample response

"One evening I was sat at home watching television when I heard my next door neighbours smoke alarm sounding. This is not an unusual occurrence as she is always setting off the alarm whilst cooking. However, on this occasion, something was different as the alarm did not normally sound so late at night. I got up out of my chair and went to see if she was OK. She is a vulnerable, elderly lady and I always look out for her whenever possible.

When I arrived next door I peered through the window and noticed my neighbour sat asleep on the chair in the front room. Wisps of smoke were coming from kitchen so I knew that she was in trouble. I immediately ran back into my house and dialled 999 calmly. I asked for the Fire Service and the Ambulance Service and explained that a person was stuck inside the house with a fire burning in the kitchen. I provided the call operator as much information as possible including landmarks close to our road to make it easier for the Fire Service to find. As soon as I got off the phone I immediately went round the back of my house to climb over the fence. Mrs Watson, my neighbour, usually leaves her back door unlocked until she goes to bed. I climbed over the fence and tried the door handle. Thankfully the door opened. I entered into the kitchen and turned off the gas heat which was burning dried up soup. I then ran to the front room, woke up Mrs Watson and carried her carefully through the front door, as this was the nearest exit.

I then sat Mrs Watson down on the pavement outside and placed my coat around her. It wasn't long before the Fire Service arrived and they took over from then on in. I gave them all of the details relating to the incident and informed them of my actions when in the kitchen."

Now take the time to use the blank space on the following page to prepare your own response to this question.

QUESTION 2

Tell me about a time when you helped someone who was distressed or in need of support?

The next PQA deals with commitment to excellence. This will show how you would display a conscientious and proactive approach to work to achieve and maintain excellent standards.

Q3. *Tell me about a time when you had to follow clear instructions or rules in order to complete a task?*

How to structure your response:

- What was the work you were doing?
- What were the rules or instructions that you had to follow?
- What did you do to complete the work as directed?
- What was the result?
- How did you feel about completing the task in this way?

Strong response

The Fire Service strives for excellence in everything it does. Therefore it is crucial that you provide a response that demonstrates you too can deliver excellence and maintain high standards. Try to think of a situation, either at work or otherwise, where you have achieved this. Make your response specific in nature. If you have had to follow specific instructions, rules or procedures then this is a good thing to tell the panel.

Weak response

Weak responses are generic in nature and usually focus on a candidate's own views on how a task should be achieved, rather than in line with a company's or organisation's policies and procedures. The candidate will display a lack of motivation in relation to following clear instructions or rules.

Sample response

"I am currently working as a sales assistant for a well-known retailer. I recently achieved a temporary promotion and part of that role includes carrying out pre-opening checks. I am required to get to work 60 minutes before opening time and carry out a comprehensive routine check. The work includes checking that all fire exits are unlocked, testing the fire alarm, assessing the current stock levels to make sure we have enough for the day's trade, turning on power and heating, checking the tills are stocked with cash, carrying out a risk assessment, briefing staff on safety

hazards, briefing staff on the requirements for the day and liaising with the shopping centre manager.

It is important that I follow the rules and instructions carefully because if I miss any of them off, the day's trading will not run smoothly and there could also be safety implications.

In order to make sure that I follow the instructions carefully I always make sure that I arrive at work with plenty of time to spare. This ensures that I leave plenty of time for any last minute hiccups. I also follow a self-made checklist which I carry around with me on a clip board. Once I have completed a task, I tick it off and write down any relevant notes that will help me to brief my staff. I always feel good about the manner in which I carry out my duties. I am an organised person and I take great pride in carrying out my duties both diligently and professionally."

Now take the time to use the blank space on the following page to prepare your own response to this question.

QUESTION 3

Tell me about a time when you had to follow clear instructions or rules in order to complete a task?

Q4. *Tell me about a time when you sought to improve the way that you or others do things?*

How to structure your answer:

- What was the improvement that you made?
- What prompted this change?
- What did you personally do to ensure that the change was successful?
- What was the result?

Strong response

Part of the firefighter's role includes continuous improvement and being able to adapt to change. Stronger performing candidates are able to provide a response that demonstrates a voluntary willingness to improve or change the way they do things.

Weak response

Candidates who are unable to identify where improvements are needed will generally provide weak responses. Once again the response will be generic in nature and lack any substance or specific evidence.

Sample response

"I currently work as a telecommunications engineer and I have been doing this job for nine years now. I am very well qualified and can carry out the tasks that form part of my job description both professionally and competently. However, with the introduction of wireless telecommunications I started to feel a little bit out of my depth. Wireless telecommunications basically provides telephone, Internet, data, and other services to customers through the transmission of signals over networks of radio towers. The signals are transmitted through an antenna directly to customers, who use devices, such as mobile phones and mobile computers, to receive, interpret, and send information. I knew very little about this section of the industry and decided to ask my line manager for an appraisal. During the appraisal I raised my concerns about my lack of knowledge in this area and she agreed to my request for continuing professional training in this important area.

I was soon booked on a training course which was modular in nature and took seven weeks to complete. During the training I personally ensured that I studied hard, followed the curriculum and checked with the course tutor periodically to assess my performance. At the end of the training I received a distinction for my efforts. I now felt more comfortable in my role at work and I also started to apply for different positions within the company that involved wireless technology. For the last 6 months I have been working in the wireless telecommunications research department for my company and have excelled in this new area of expertise."

Now take the time to use the blank space on the following page to prepare your own response to this question.

QUESTION 4

Tell me about a time when you sought to improve the way that you or others do things?

The next PQA deals with commitment to development

This will show your commitment and ability to develop yourself and others:

Q5. *Tell me about a time when you have taken it upon yourself to learn a new skill or develop an existing one?*

How to structure your response:

- What skill did you learn or develop?
- What prompted this development?
- When did this learning or development occur or take place?
- How did you go about learning or developing this skill?
- What was the result?
- How has this skill helped you since then?

Strong response

Firefighters are required to learn new skills every week. They will attend on-going training courses and they will also read up on new procedures and policies. In order to maintain a high level of professionalism, firefighters must be committed to continuous development. Try to think of an occasion when you have learnt a new skill, or where you have taken it upon yourself to develop your knowledge or experience in a particular subject. Follow the above structure format to create a strong response.

Weak response

Those candidates who have taken on any new development or learning will be unable to provide a strong response. They will provide a response where they were told to learn a new skill, rather than taking it upon themselves. There will be no structure to their learning or development and they will display a lack of motivation when learning.

Sample response

"Although I am in my late thirties I had always wanted to learn to play the guitar. It is something that I have wanted to do for many years, but have never had the time to learn until recently. One day I was watching a band play with my wife at my local pub and decided there and then that I would make it my mission to learn to play competently. The following day I went onto the internet and searched for a good guitar tutor in my local

area. Luckily, I managed to find one within my town who had a very good reputation for teaching. I immediately booked a block of lessons and started my first one within a week. My development in the use of playing the guitar progressed rapidly and I soon achieved grade 1 standard. Every night of the week I would dedicate at least 30 minutes of time to my learning, in addition to my one hour weekly lesson. I soon found that I was progressing through the grades quickly, which was due to my level of learning commitment and a desire to become competent in playing the instrument.

I recently achieved level 4 and I am now working to level 5 standard. I am also now playing in a local band and the opportunities for me, both musically and socially, have increased tenfold since learning to play. In addition to this, learning to play the guitar has improved my concentration levels and my patience."

Now take the time to use the blank space on the following page to prepare your own response to this question.

QUESTION 5

Tell me about a time when you have taken it upon yourself to learn a new skill or develop an existing one?

Q6. *Tell me about a time when you changed how you did something in response to feedback from someone else?*

How to structure your response:

- What did you need to develop?
- What feedback did you receive and from whom?
- What steps did you take to improve yourself or someone else?
- What did you specifically say or do?
- What was the result?

Strong response

Firefighters receive feedback from their supervisory managers on a regular basis. In their quest to continually improve, the Fire Service will invest time, finances and resources into your development. Part of the learning process includes being able to accept feedback and also being able to improve as a result of it. Strong performing candidates will be able to provide a specific example of where they have taken feedback from an employer or otherwise, and used it to improve themselves.

Weak response

Those candidates who are unable to accept feedback from others and change as a result will generally provide a weak response to this type of question. They will fail to grasp the importance of feedback and in particular where it lies in relation to continuous improvement. Their response will be generic in nature and there will be no real substance or detail to their answer.

Sample response

"During my last appraisal, my line manager identified that I needed to improve in a specific area. I work as a call handler for a large independent communications company. Part of my role involves answering a specific number of calls per hour. If I do not reach my target then this does not allow the company to meet its standards. I found that I was falling behind on the number of calls answered and this was identified during the appraisal. I needed to develop my skills in the manner in which I handled the call. My line manager played back a number of recorded calls that I had dealt with and it was apparent that I was taking too long speaking to the customer about issues that were irrelevant to the call itself. Because I am conscien-

tious and caring person I found myself asking the customer how they were and what kind of day they were having. Despite the customers being more than pleased with level of customer care, this approach was not helping the company and therefore I needed to change my approach. I immediately took on-board the comments of my line manager and also took up the offer of development and call handling training. After the training, which took two weeks to complete, I was meeting my targets with ease. This in turn helped the company to reach it's call handling targets."

Now take the time to use the blank space on the following page to prepare your own response to this question.

QUESTION 6

Tell me about a time when you changed how you did something in response to feedback from someone else?

The next PQA is the last and it deals with commitment to diversity and integrity. This will show how you understand and respect other people's values and how you would adopt a fair and ethical approach to others. Diversity means the differences that exist between people, such as gender, age, ethnic background, religion, social background etc.

Q7. *Describe a time when you have helped to support diversity in a team, school, college or organisation.*

How to structure your response:

- What was the situation?
- What prompted the situation?
- What were the diversity issues?
- What steps did you take to support others from diverse backgrounds?
- What specifically did you say or do?
- What was the result?

Strong response

This type of question is difficult to respond to, especially if you have little or no experience in this area. However, strong performing candidates will be able to provide clear details and examples of where they have supported diversity in a given situation. Their response will be specific in nature and it will clearly indicate to the panel that they are serious about this important subject.

Weak response

Weak responses are generic in nature and they fail to answer the question that is being asked. Many candidates are unable to provide a specific response to this type of question.

Sample response

"Whilst working in an office I noticed that a new member had joined our team. The lady was confined to a wheelchair and for some strange reason my boss decided to put her in an office which was a fair distance from the exit, toilet and kitchen facilities. I immediately picked up on this and decided that something needed to be done in order to help make her life a bit easier. Although the office shape and design met current legal requirements for access, I felt it was unfeasible to expect the new member of the team to

have such a struggle to get to these everyday facilities. I started off by approaching my boss and asking him if I could make a suggestion. I started to put my case over to him and explained that I would be more than happy to offer the lady my office, as it was much closer to the facilities. My boss agreed that it was a good idea and he felt rather guilty that he had not even considered the issue in the first place.

I then went over to my new colleague and introduced myself. I quietly asked her if she would like to swap offices with me and explained that the facilities would be a lot closer for her if she took up my offer. I also apologised for her being put so far away from them in the first place. She thanked me for my consideration and took me up on my offer. I then assembled a small team of workers who assisted me in moving desks and computers. The lady soon settled into her new office where she had much better access to the exit and the facilities.

Finally, I suggested to my boss that we should all attend a company diversity awareness course to raise awareness of these important issues. He agreed and we all attended the course over a period of four weeks."

Now take the time to use the blank space on the following page to prepare your own response to this question.

QUESTION 7

Describe a time when you have helped to support diversity in a team, school, college or organisation.

Q8. *Tell me about a time when you noticed a member of your team or group behaving in a manner which was inconsistent with the teams, groups, or organisation's values?*

How to structure your answer:

- What was the situation?
- How was the behaviour inconsistent with the team's or organisation's values?
- Why were the colleagues behaving in that way?
- What did you say or do when you noticed this behaviour?
- What difficulties did you face?
- What was the result?

Strong response

Firefighters need to have the confidence and ability to challenge unacceptable behaviour whilst at work. In order to understand what unacceptable behaviour is, you first need to know what the values of the organisation are. Candidates who provide a strong response will have a clear understanding of an organisation's values and also how to tackle unacceptable behaviour in the correct manner.

Weak response

Weak responses are generally where a candidate is unaware of the importance of an organisation's values and how they impact on the needs of a team or group. They will not have the confidence to challenge inappropriate behaviour and they will turn a blind eye whenever possible. Their response will lack structure and it will be generic in nature.

Sample response

"Whilst working as a sales person for my previous employer, I was serving a lady who was from an ethnic background. I was helping her to choose a gift for her son's 7th birthday when a group of four youths entered the shop and began looking around at the goods we had for sale. For some strange reason they began to make racist jokes and comments to the lady. I was naturally offended by the comments and was concerned for the lady to whom these comments were directed. Any form of bullying and harassment is not welcome in any situation and I was determined to stop it immediately and protect the lady from any more harm.

The lady was clearly upset by their actions and I too found them both offensive and insensitive. I decided to take immediate action and stood between the lady and the youths to try to protect her from any more verbal abuse or comments. I told them in a calm manner that their comments were not welcome and would not be tolerated. I then called over my manager for assistance and asked him to call the police before asking the four youths to leave the shop. I wanted to diffuse the situation as soon as possible, being constantly aware of the lady's feelings. I was confident that the shop's CCTV cameras would have picked up the four offending youths and that the police would be able to deal with the situation. After the youths had left the shop I sat the lady down and made her a cup of tea whilst we waited for the police to arrive. I did everything that I could to support and comfort the lady and told her that I would be prepared to act as a witness to the bullying and harassment that I had just witnessed.

I believe the people acted as they did because of a lack of understanding, education and awareness. Unless people are educated and understand why these comments are not acceptable then they are not open to change. They behave in this manner because they are unaware of how dangerous their comments and actions are. They believe it is socially acceptable to act this way when it certainly isn't.

I also feel strongly that if I had not acted and challenged the behaviour the consequences would be numerous. To begin with I would have been condoning this type of behaviour and missing an opportunity to let the offenders know that their actions are wrong (educating them). I would have also been letting the lady down, which would have in turn made her feel frightened, hurt and not supported. We all have the opportunity to help stop discriminatory behaviour and providing we ourselves are not in any physical danger then we should take positive action to stop it."

Now take the time to use the blank space on the following page to prepare your own response to this question.

QUESTION 8

Tell me about a time when you noticed a member of your team or group behaving in a manner which was inconsistent with the teams, groups, or organisation's values?

Q9. *Tell me about a time when you made a mistake that had a knock-on effect on other people?*

How to structure your answer:

- What were the circumstances?
- What was the mistake?
- How did the mistake affect others?
- What did you do or say when you noticed your mistake?
- What was the outcome?
- What would you do differently next time?

Strong response

We all make mistakes from time to time but what is important is what you do after the mistake. Strong performing candidates will provide honest details about the mistake they made and also how they improved as a result of it. They will also provide details of how they reflected on their performance and how they evaluated their actions with a view to improve.

Weak response

Candidates who provide a weak response will state that they are perfect and that they never make mistakes. Their response will not provide an example of where a genuine mistake has been made and what they did specifically to learn from it.

Sample response

"I can remember being responsible for carrying out a stock take at the shop I was working at. It was late on a Friday afternoon and I had volunteered to carry out the task after closing hours as no other member of staff was available. I can remember rushing the task because I was due to go out with my girlfriend for a meal that evening. This was a wrong way to approach the task as it resulted in my figures being out for a number of items. The mistake had a knock-on effect as, when the shop was opened on Monday morning, there was insufficient stock available for the week ahead. This resulted in a number of items being unavailable, which did not look good for the company or our valued customers. The mistake was made aware to me by another member of staff on the Monday morning. I immediately went to my line manager's office and informed her of the mistake. I apologised

unreservedly and offered to drive to the outlet warehouse in order to pick up the extra stock needed. She stated that this was not permitted as I was required to help out on the shop floor. I felt really bad about my mistake but was determined to learn from it and help out in any way possible to resolve it.

I asked my line manager if it would be possible for me to stay behind again on the forthcoming Friday to shadow the member of staff who would be carrying out the stock take. This would enable me to carry out the stock take again in a controlled and supervised manner so that no mistakes would be made. She agreed to this and thanked me for my mature approach to the situation."

Now take the time to use the blank space on the following page to prepare your own response to this question.

QUESTION 9

Tell me about a time when you made a mistake that had a knock-on effect on other people?

TIPS FOR ANSWERING THE PQA INTERVIEW QUESTIONS

> Obtain a copy of the PQA's relevant to the role of a Firefighter. Make sure you learn them thoroughly. Then, try to predict as many possible interview questions from them. For example, if one of the PQA's is relevant to 'working with others' then it is highly likely that you will be asked a question where you are required to provide evidence of where you have successfully worked as part of a team.

> It is always best to say if you had to ask for advice or consulted someone on how you were dealing with a problem.

> Remember that the questions provided in this section are only samples based around the PQAs. You could be asked an interview question that is different to those above but the questions will always be based around the PQAs. If you prepare a number of different scenarios based around the firefighter PQAs then you won't go far wrong.

> It is always positive to say whether you would have done something differently second time around if a situation occurred again, because you felt you could have done it better.

> Be prepared for probing questions. Examples of probing question include:

i) How did that make you feel?

ii) Would you do anything differently next time?

iii) How did the other person(s) react to your actions?

iv) Why did you do it that way?

v) What did you learn from the process?

vi) Can you give me an alternative example?

> Evaluating and reflecting on your actions is very good; you are marked on this but interviewers are not allowed to ask you unfortunately. This will get you more points. Try, at some point during your responses, include the words 'evaluate' and 'reflect'.

> It is important for your sake that you choose a number of different situations.

Therefore we advise that you write down examples from your experience before you attend the interview. Always have more than one, that way you will take the pressure off yourself to remember them all.

> You are not permitted to take notes into the interview with you but it is advisable that you write down your responses during your preparation – this will help you to remember them!

GENERIC QUESTIONS BASED AROUND THE FIREFIGHTER'S ROLE AND YOUR KNOWLEDGE OF THE FIRE SERVICE

Within this section of the guide I have provided you with a number of generic questions that you may be asked during your firefighter interview. Do not rely on the questions to come up during your interview but rather use them as a preparation tool only.

It is a good idea to practise for interviews with a friend or relative. Set up a room as you'd expect to find it in an interview and get the person to ask you questions. At the end of the mock interview ask them for feedback on how they thought you performed.

Following the initial list of questions, I have supplied you with explanations to some of the questions and a sample response to show how the question might be answered. Please remember that the responses are for guidance purposes only. Your responses should be based on your own individual skills, knowledge and experience.

SAMPLE GENERIC QUESTIONS

Why have you applied to join the Fire Service and what do you have to offer?

What do you understand about the role of a firefighter?

What qualities and attributes do you think a firefighter should have?

What is Community Fire Safety?

If a member of the public asked you how to call the Fire Service, what advice would you give them?

Do you have any experience of working as a team member?

How would you change your approach when dealing with a child during an operational incident?

Do you have any experience of working within the community?

How would you react to an angry member of the public whilst attending a fire?

Have you ever had to deal with an emergency and if so what did you do and why?

What do you understand by the term diversity?

Who is the Brigade Manager/Chief Executive of this organisation?

What do you understand about the term Health and Safety?

Who is responsible for Health and Safety?

What do you understand about the term Equality and Fairness?

If you witnessed a member of your team being bullied or harassed at work, what action would you take and why?

The role of a firefighter involves working long and unsociable hours. How do you think you would cope with this?

What types of operational incidents do you think the Fire Service attends?

How do you think you would cope when dealing with casualties at incidents?

How do you deal with stress?

What qualities do you possess that would be of benefit to the Fire Service?

What are your strengths?

What are your weaknesses?

What would be your reaction if someone you were working with was acting in an offensive manner? (Racially or sexually)

What would you do if you witnessed a friend in the Fire Service stealing?

Can you tell me a time when you have made a bad decision in life and how you dealt with it?

Can you give any examples of where you have had to carry out work of a practical nature?

How do you keep yourself fit and why do you think fitness is important to the role of a firefighter?

What are the current Community Fire Safety activities of this Fire and Rescue Service?

Part of a firefighter's role is to carry out routine tasks such as cleaning of equipment and maintenance. How do you think you would cope with this element of the job?

If you overheard some information that related to a work colleague committing a criminal offence, what would your reaction be?

How many fire stations are there within this Fire and Rescue Service?

What do you understand about the term 'Integrated Risk Management Plan'? (NOTE – This can be found on the majority of Fire and Rescue websites.)

What do you know about the structure of this Fire Service?

How many fire stations are there in the county?

How would you feel about working with a gay or lesbian firefighter?

On your way to work one morning you witness a car accident. What would you do and why?

Have you ever had any experience of taking responsibility?

Do you currently have any responsibilities either at home or at work?

How do you arrive at making difficult decisions?

What experience do you have in communicating with the general public?

What do you think the important elements of communicating with different groups of people are?

How do you respond to pressure?

Do you have any questions for the panel?

SAMPLE GENERIC INTERVIEW QUESTION - NUMBER 1

Why have you applied to join the Fire Service and what do you have to offer?

This question is a common one amongst Fire Service interview panels and one that you should prepare for. The question may be asked in a variety of different forms but an explanation of why you have applied and what you have to offer is something that the panel will want to hear. When answering this type of question, remember to focus your response on the role of the firefighter and how you are suited to it.

The question may be worded in a way that asks you why you want to join their particular Fire and Rescue Service. If this is the case, you are advised to respond with an answer that reflects the positive aspects of that particular Fire Service.

The following is a sample response to this type of question. Once you have read the response, use it as a guide to help you construct your own answer.

SAMPLE GENERIC INTERVIEW QUESTION RESPONSE - NUMBER 1

Why have you applied to join the Fire Service and what do you have to offer?

"Joining the Fire Service is something that I have wanted to do for many years now and I have been preparing for the selection process for some time. I believe that I would enjoy working in a community-focused service where the priority is helping other people, both through preventative work and reactive operational work.

Having researched the Fire Service extensively, I have been impressed and attracted to the shift from predominantly reactive work to prevention work in terms of Community Fire Safety. I enjoy learning new skills and keeping up to date with procedures and policies, and believe the Fire Service would be a career that I would be very much suited to.

I am a caring person who is comfortable working with people, regardless of their background, age, sexuality or gender. I keep myself physically fit and understand that this is an important aspect of the firefighter's role. I want to become a firefighter with this Fire and Rescue Service because I believe I can make a positive difference to the team. I am enthusiastic, motivated, focused and driven and would love to work for a professional service that has such high standards."

SAMPLE GENERIC INTERVIEW QUESTION - NUMBER 2

What do you understand about the role of a firefighter?

Once again, this is a common question used by UK Fire and Rescue Services. The question is used to test your knowledge of the firefighter's role.

Many people will just respond with an answer that solely relates to the operational aspect of the role. Such a response will not be adequate to attract high marks. Remember to make reference to the Personal Qualities and Attributes (PQAs) that are applicable to the firefighter's role.

Focus your response around the PQAs and the diverse role of the modern-day firefighter. Take a look at the following sample response before creating your own based on your own skills, knowledge and experience.

SAMPLE GENERIC INTERVIEW QUESTION RESPONSE - NUMBER 2

What do you understand about the role of a firefighter?

"The two main areas of the firefighter's role are responding to incidents as and when they arise, and also Community Fire Safety work, which involves looking for ways to reduce incidents. I understand that the role is based around the Personal Qualities and Attributes for the firefighter. This includes a commitment to diversity and integrity and also being open to change.

The Fire Service is a continually changing and improving service and it is important that firefighters embrace this. A firefighter must have the confidence and resilience to deal with highly stressful situations and be capable of working with all people, regardless of age, sexuality, gender or background.

Because firefighters carry out so much community work, they have to be effective communicators and be capable of solving problems as and when they arise. Also, because the Fire Service is a customer-focused service, the firefighter needs to be committed to delivering an excellent service. The role also includes training and maintaining competence through the Integrated Personal Development System.

Firefighters are also constantly looking for ways to reduce the risk of fire and providing advice to the public upon request. Maintaining an operational readiness is also key to the role, making sure that all equipment is serviced and ready for use. Having the ability to work as an effective team member is essential and having a good knowledge of Health and Safety is also important."

SAMPLE GENERIC INTERVIEW QUESTION - NUMBER 3

What is Community Safety?

If you are serious about becoming a firefighter then you should have a good knowledge of Community Safety.

Before preparing your answer to this question I recommend that you visit the website of the Fire Service you are applying to join so that you can get a feel for what approach they are taking in relation to CFS. You may also wish to visit the Government's own Community Fire Safety site to get some more useful tips about this important subject.

Now take a look at the following sample response.

SAMPLE GENERIC INTERVIEW QUESTION RESPONSE - NUMBER 3

What is Community Safety?

"Community safety is one of the core elements of the firefighter's role.

It is about informing and educating the public with safety information that will help them to reduce the risk of fire in the home to nearly zero. It covers many different areas ranging from information relating to smoke alarms, home fire safety checks, electrical fire safety and cooking safety to name but a few.

I am aware that the Fire Service is constantly looking for ways to reduce fire deaths and injuries in the home through its effective Community Safety reduction strategies.

Community Safety is also about working with other agencies including the Police and Social Services to establish ways of making the community safer together as opposed to working in isolation.

I visited your website and noticed that you have been working with Help the Aged to provide smoke alarms for the elderly, which is a good example of agencies working together to help save lives."

SAMPLE GENERIC INTERVIEW QUESTION - NUMBER 4

If a member of the public asked you how to call the Fire Service, what advice would you give them?

This question is not a common one but there have been occasions when it has been asked during the firefighter interview. If you are going to be a firefighter then you certainly should know how to call the Fire Service in the event of an emergency.

The answer is a simple one and the following is a sample response to help you.

SAMPLE GENERIC INTERVIEW QUESTION RESPONSE - NUMBER 4

If a member of the public asked you how to call the Fire Service, what advice would you give them?

"I would tell them to dial 999 using the nearest available working phone.

I'd also inform them that they can use their mobile phone to dial 999 even if they do not have any credit available.

I would tell them that they would be connected to a central call handling centre where they will be asked which service they require.

I would tell them that they must ask for the Fire Service. Once they are through to the Fire Service operator they will be asked a series of important questions. I would tell them to listen carefully to the Fire Service operator and answer all the questions carefully and accurately. It is important that they remain calm when making the call so that the operator can obtain all of the information.

I would tell them that the type of questions they will be asked are:

-What the emergency is (e.g. fire, car crash, flood, person trapped etc);

-Where it is (e.g. full address if known, name of road, prominent landmark);

-How many people are involved, if any?

-Any special problems/hazards that they need to know about.

I would finally inform them that it is important to only use the 999 service when it is genuinely needed and that hoax calls should never be made."

SAMPLE GENERIC INTERVIEW QUESTION - NUMBER 5

Do you have any experience of working as a team member?

The ability to work effectively in a team is an extremely important aspect of the firefighter's role. Not only will you be spending a great deal of time together at work, you will also depend on your colleagues during highly dangerous and stressful incidents. Therefore it is important that you can demonstrate you have the ability to work as an effective team member.

When responding to this type of question, try to think of occasions when you have been part of a team and achieved a common goal.

Maybe you are already involved in team sports playing hockey, rugby or football? You may also find that you have experience of working as a team member through work. If you have no or very little experience of working as a team member then try to get some before you apply to the Fire Service. After all, teamwork is an important aspect of the role.

Now take a look at the following sample response.

SAMPLE GENERIC INTERVIEW QUESTION RESPONSE - NUMBER 5

Do you have any experience of working as a team member?

"Yes I have many years experience of working in a team environment.

To begin with, I have been playing hockey for my local team for the last 3 years. We worked really hard together improving our skills over the course of last season and we managed to win the league.

I am also very much involved in teamwork in my current job. I work as a nurse at the local hospital and in order for the ward to function correctly we must work effectively as a team. My job is to check all of the patients at the beginning of my shift and also make sure that we have enough medical supplies to last the duration. It is then my responsibility to inform the ward sister that the checks have been carried out. She will then obtain more supplies if we need them.

We have to work very closely together for many hours and we all pull together whenever the going gets tough. I enjoy working in a team environment and feel comfortable whilst working under pressure."

SAMPLE GENERIC INTERVIEW QUESTION - NUMBER 6

Do you have any experience of working within the community?

Because the firefighter's role is very much community based, the Fire Service want to know that you are able to work with people from all backgrounds.

Many people would not feel comfortable working in the community but as a firefighter it is an essential part of your role. Firefighters have very good reputations for being caring, helpful and considerate people who will help out wherever possible. Whilst working as a firefighter you will be out in the community promoting Fire Safety, visiting people's homes to offer fire safety advice and fitting smoke alarms etc. Therefore it is important that you can provide examples of where you have already carried out some form of community work.

Community work can involve many different things ranging from Neighbourhood Watch to charity work or voluntary work.

Take a look at the following sample response.

SAMPLE GENERIC INTERVIEW QUESTION RESPONSE - NUMBER 6

Do you have any experience of working within the community?

"I recently organised a charity boot fair at my local school. This was to try to raise money for a nearby hospital that wanted to buy some new medical equipment. I worked with a number of different people in the community to get the event off the ground.

I worked closely with the local school and advertised the boot fair in the local paper to try to generate some interest. I contacted local community groups such as Neighbourhood Watch to try to promote the event, which worked very well. The boot fair was attended by over 500 people and we managed to raise over £750 for the good cause. I wouldn't have been able to arrange the event without working closely with different people and groups from within the community.

The event was a great success and I plan to arrange another one next year."

SAMPLE GENERIC INTERVIEW QUESTION - NUMBER 7

What do you understand by the term diversity?

You are almost guaranteed to be asked a question that relates to diversity and working with people from different cultures and backgrounds.

One of the Personal Qualities and Attributes of a firefighter is the ability to work with others. Over the last few years, Senior Fire Service stakeholders, in collaboration with the Government, have taken up the challenge to work towards a more diverse workforce. Therefore, an understanding of what diversity means, and how important it is to the Fire Service, is crucial if you are to become a firefighter. This particular question is designed to see if you understand what the term diversity means in relation to the Fire Service.

Take a look at the following sample response to this question.

SAMPLE GENERIC INTERVIEW QUESTION RESPONSE - NUMBER 7

What do you understand by the term diversity?

"The term diversity means different and varied.

For example, if the Fire Service has a diverse workforce, it means that the people in that workforce are from different backgrounds, cultures and genders. The community in which we live is extremely diverse. Therefore, it is important that the Fire Service represents the community in which it serves so that a high level of service can be maintained. This gives the public more confidence in the Fire Service.

There are also other added benefits of a diverse workforce. It enables the Fire Service to reach every part of the community and provide Fire Safety advice to everybody as opposed to just certain individual groups of people."

SAMPLE GENERIC INTERVIEW QUESTION - NUMBER 8

What do you understand about the term Health and Safety and who is responsible for it?

Health and Safety plays a very important part in the firefighter's working day.

As a firefighter you will be acutely aware of Health and Safety and how it affects you and your colleagues. Health and Safety is the responsibility of everybody at work. You are responsible for the safety of yourself and for the safety of everybody else. Health and Safety within the Fire Service is governed by the Health and Safety at Work Act 1974 and the Management of Health and Safety at Work Regulations 1999.

Make sure you are aware of the term 'risk assessment' and what it means to the operational firefighter.

The following is a sample response to this type of question.

SAMPLE GENERIC INTERVIEW QUESTION RESPONSE - NUMBER 8

What do you understand about the term Health and Safety and who is responsible for it?

"Everybody is responsible for Health and Safety at work. Health and Safety is governed by the Health and safety at Work Act 1974 and the Management of Health and Safety at Work Regulations 1999. Firefighters are responsible for the safety of themselves and the safety of each other.

Health and Safety is all about staying safe and promoting good working practices. In the Fire Service this means making sure that all protective clothing is usable and in good working order, checking that equipment and machinery is serviceable and carrying out risk assessments when required. It also includes simple things like making sure warning signs are placed out after the fire station bays have been cleaned.

It applies both when at fires and incidents and also when carrying out duties around the station. Health and Safety should be at the forefront of everybody's minds when at work."

SAMPLE GENERIC INTERVIEW QUESTION - NUMBER 9

What do you understand about the term Equality and Fairness?

Treating everybody with respect and dignity is important in everyday life. Treat others how you would expect to be treated regardless of their age, gender, sexual orientation or cultural background.

If you are not capable of treating people with respect and dignity then the Fire Service is not for you!

A question based on this subject is likely to come up during the interview and it relates to the PQA 'working with others'.

The following is a sample response to this question.

SAMPLE GENERIC INTERVIEW QUESTION RESPONSE - NUMBER 9

What do you understand about the term Equality and Fairness?

"Equality and Fairness is about treating people with dignity and respect and without discrimination. Unfair discrimination in employment is wrong. It is bad for the individuals who are denied jobs or who suffer victimisation or harassment because of prejudice. I understand that within the Fire Service it is the responsibility of everyone to uphold the principles and policies of the organisation in relation to Equality and Fairness.

Discrimination or unacceptable behaviour of any sort is not tolerated and nor should it be. Not only is it important to apply these principles whilst working with colleagues in the Fire Service but it also applies when serving the public."

SAMPLE GENERIC INTERVIEW QUESTION - NUMBER 10

If you witnessed a member of your team being bullied or harassed at work, what action would you take and why?

There is only one answer to this question and that is that you would take action to stop it, providing it was safe to do so. Bullying or harassment of any kind must not be tolerated. The second part of the question is just as important. They are asking you why you would take this particular action.

Before you prepare your answer to this question think carefully about what action you would take if somebody was being bullied or harassed. Taking action can mean a number of different things ranging from reporting the incident to your manager, through to intervention.

Whatever answer you give it is important that you are honest and tell the truth about how you would respond to such a situation. Take a look at the following sample response to this question.

SAMPLE GENERIC INTERVIEW QUESTION RESPONSE - NUMBER 10

If you witnessed a member of your team being bullied or harassed at work, what action would you take and why?

"I would stop it immediately if it was safe to do so. This type of behaviour is totally unacceptable and must not be tolerated in the workplace.

The reason why I would take action is because if I didn't, then I would effectively be condoning the bullying or harassment. The type of action I would take would very much depend on the circumstances. In most cases I would intervene at the time of the incident and ask the person to stop the bullying or harassment.

If the incident were very serious, then I would report it to my manager so that further action could be taken. Whatever the situation was, I would definitely take steps to stop it from happening. I believe that I would also have a duty under Fire Service policy to take action to stop bullying and harassment."

SAMPLE INTERVIEW QUESTION - NUMBER 11

What would be your reaction if someone you were working with was acting in an offensive manner? (Racially or sexually)

This kind of behaviour is not tolerated and therefore you should be asking the person to stop acting in this offensive manner.

This type of behaviour includes any form of racial or sexual jokes and again, these are not tolerated within the UK Fire Service. The Fire Service has strict policies in relation to this kind of unacceptable behaviour and it will not be tolerated.

When responding to questions of this nature, you should say that you would take steps to stop the person from acting in this manner, either through intervention or reporting. However, only ever say these words if you actually mean it. Do not lie.

Take a look at the following sample response.

SAMPLE GENERIC INTERVIEW QUESTION RESPONSE - NUMBER 11

What would be your reaction if someone you were working with was acting in an offensive manner? (Racially or sexually)

"I would ask the person to stop. That would be my first action. This kind of behaviour is not acceptable. If I was to ignore it then I would be just as bad as the person who was carrying out the act.

I would then inform my line manager about the behaviour so that he/she could decide if any further action needed to be taken."

SAMPLE GENERIC INTERVIEW QUESTION - NUMBER 12

What qualities do you possess that would be of benefit to the Fire Service?

Questions based around 'qualities' are designed to assess that you have the right 'Personal Qualities and Attributes' to become a firefighter. Therefore, your response should be predominantly based around these. However, it is still important to remember that a firefighter has many other qualities aside from these and you should try to include them too where possible.

Take a look at the following sample response before constructing your own.

SAMPLE GENERIC INTERVIEW QUESTION RESPONSE - NUMBER 12

What qualities do you possess that would be of benefit to the Fire Service?

"To begin with I am a physically fit person who takes pride in my appearance. I am punctual, reliable and can be trusted to carry out a task on time and to a high standard.

I am a very good team worker and work well with other people. I can also be relied upon to work on my own unsupervised wherever necessary. I have good problem solving skills, which have been obtained through my practical working background.

I am a confident person who is always looking to improve, learn and develop new skills. I am extremely adaptable and I am always open to new ideas. I view change as a positive thing and enjoy the challenges that this brings.

I have very good communication skills, both written and verbal, and I am committed to working hard for my employer, whoever that may be."

SAMPLE GENERIC INTERVIEW QUESTION - NUMBER 13

Can you tell me a time when you have made a bad decision in your life and how you dealt with it?

Not a common question, but one that has been asked on a number of occasions. This is quite a difficult one to answer as we have all made bad decisions in our life and anybody who says that they haven't is probably not telling the truth.

However, the response that you give is important. You need to provide an example that doesn't put you across in a bad light.

The second part of the question is just as important. How you deal with situations or mistakes that you have made in your life is a good indication of your character and maturity.

Take a look at the example we have provided on the following page before preparing your own based on your individual experiences.

SAMPLE INTERVIEW QUESTION RESPONSE - NUMBER 13

Can you tell me a time when you have made a bad decision in your life and how you dealt with it?

"Yes. About 5 years ago I was working at a local shop stacking shelves and working behind the counter serving customers. I was working very hard for six days a week for a very low salary.

I was unhappy in the job because of the poor wages and although I liked my boss I felt that it was not worth working so hard for such poor wages, so I decided to leave.

I then found myself in a position having to find a new job without any money. I was studying at night school to gain an educational qualification and therefore needed some money for books and study material etc. Leaving the job without having another one to go to was a bad decision on my part. I should have waited until I found another job before leaving.

However, I didn't get down about it. I immediately started looking for further employment and after 2 weeks I managed to get a job as a customer sales advisor for a local car retail firm, which I enjoyed very much.

If I make a bad decision in life I always try to take steps to rectify it but more importantly learn from it. I believe it is important to evaluate and reflect on the important decisions we make in life."

SAMPLE GENERIC INTERVIEW QUESTION - NUMBER 14

What would you do if you witnessed a friend in the Fire Service stealing?

There is only one answer to this question and that is to report the person who is stealing. Firefighters are trusted in the community and relied upon to be open, honest and reliable.

You will find that you will attend many incidents where you will be going in people's homes, vehicles or property and the Fire Service trusts you to act in accordance with their policies and procedures.

The following is a sample response to this question.

SAMPLE GENERIC INTERVIEW QUESTION RESPONSE - NUMBER 14

What would you do if you witnessed a friend in the Fire Service stealing?

"I would report the incident to my line manager. Of course it would be a difficult decision to make because that person is my friend and colleague. However, I would not like it if a firefighter was stealing from my house or property and therefore it should be stopped immediately. I would also say that any person who steals from others is no friend of mine.

I would choose the appropriate time to inform my line manager about the incident. I understand that firefighters are trusted people in society and anything that does harm to that reputation should be challenged."

SAMPLE GENERIC INTERVIEW QUESTION - NUMBER 15

What are the current Community Fire Safety activities of this Fire and Rescue Service?

A question that relates to Community Fire Safety is more than likely to come up during the interview. Remember to read the Community Fire Safety section of this guide to get a basic understanding of the type of advice the UK's firefighters are giving the public.

Make sure that you visit the website of the Fire and Rescue Service you are applying for. On this you will find some useful information that relates to CFS and what the Fire Service is doing to make the community a safer place.

Take a look at the following sample response to this question before constructing your own.

SAMPLE GENERIC INTERVIEW QUESTION RESPONSE - NUMBER 15

What are the current Community Fire Safety activities of this Fire and Rescue Service?

"Having visited your website and spoken to serving firefighters at my local fire station, I understand that you are currently carrying out a number of Home Fire Safety Checks in people's homes to try to reduce the number of accidental house fires and injuries.

I also noticed that you are working closely with other agencies such as the Police to try to reduce the number of deliberate vehicle fires that are occurring. I have also visited the Government's website to see what is happening on a national scale.

I particularly liked the 'checklist' that you provide on your website so that members of the public can make sure that they are doing all they can to keep themselves safe from fire."

SAMPLE GENERIC INTERVIEW QUESTION - NUMBER 16

How do you keep yourself fit and why do you think fitness is important to the role of a firefighter?

Fitness is an important element of the firefighter's role.

As an operational firefighter you will have a duty to keep yourself physically fit and active. This question is quite simple to answer providing that you do actually carry out some form of physical exercise. If you do not then now is a good time to start. If you play a team sport then again this would be an advantage.

Eating properly is also key to maintaining a healthy lifestyle. Whilst you are unlikely to be asked questions about your diet, you will find that you feel a whole lot better about yourself if you eat properly.

The following is a sample response to this question, based on a person who lives a healthy lifestyle and keeps physically fit.

SAMPLE GENERIC INTERVIEW QUESTION RESPONSE - NUMBER 16

How do you keep yourself fit and why do you think fitness is important to the role of a firefighter?

"Yes, I keep myself fit and active and it is an important part of my life. I go swimming 3 times a week and swim 30 lengths every time I go.

I also go jogging 2 times a week to break up the routine and mix in some light weight work at the gym every now and again.

I also play football for my local Sunday team, which involves one practice session every fortnight. Plus I ensure that I eat a proper diet, which helps to keep me feeling confident and healthy.

Yes, I think that fitness is vital to the role of a firefighter. The role involves working unsociable hours and I understand that it can be highly stressful at times. Coupled with the fact that the work can be physically demanding, it is important that firefighters stay fit and healthy so that they can perform to their peak and cope with the demands of the job."

SAMPLE GENERIC INTERVIEW QUESTION - NUMBER 17

Have you ever had any experience of taking responsibility?

Firefighters must be responsible people.

You are responsible for your own safety and the safety of your work colleagues. You are responsible for looking after your personal protective equipment and for turning up for work on time and prepared. A question based around responsibility is likely during the firefighter interview so we recommend that you take the time to think up an example of where you have had to take responsibility. This can either be in your working life, social or home life.

Take a look at the following sample response before creating your own response based on your own individual circumstances.

SAMPLE GENERIC INTERVIEW QUESTION RESPONSE - NUMBER 17

Have you ever had any experience of taking responsibility?

"Yes, I am a single parent who has a great deal of responsibility every day of the week. I look after my two children, making sure they get to school every morning on time, before going to work myself.

My current day job is as a customer sales representative for a large retail outlet. Within this role I also have many responsibilities, which include the supervision of stocktaking and ensuring that all goods are ordered on time.

After I have finished work I pick up my children from the childcare centre before going home and helping them with their homework. I basically have responsibility for the total running of the house and making sure that everything runs smoothly.

I enjoy the challenge of responsibility and understand that it is an important aspect of the firefighter's role."

SAMPLE GENERIC INTERVIEW QUESTION - NUMBER 18

What do you think the important elements of communicating with different groups of people are?

A question based around your communication skills is likely to appear during the firefighter interview. This question is designed to assess two parts of the Firefighter Personal Qualities and Attributes – working with others and effective communication skills.

Remember that firefighters have to be good communicators and be capable of working with people from every part of the community.

The following is a sample response to this question to help you structure your own.

SAMPLE GENERIC INTERVIEW QUESTION RESPONSE - NUMBER 18

What do you think the important elements of communicating with different groups of people are?

"Effective communication skills are an integral part of the firefighter personal qualities and attributes. One of the main elements is respect. Being respectful of people's backgrounds and trying to appreciate how they feel about things, particularly the Fire Service, is important so that good relationships can be built.

Listening effectively is also another important aspect of good communication.

Listening to what people say and getting feedback is important so that improvements can be made.

When communicating it is vital that firefighters create an approachable and positive image so that trust can be built. Asking questions is important too so that you can understand what the different groups' needs are."

SAMPLE GENERIC INTERVIEW QUESTION - NUMBER 19

How would you change your attitude/approach when talking to a child who was injured at an incident?

TIP: this is a very good question to prepare for. Not many candidates will be prepared for this, simply because it is unlikely to come up during the interview. However, if it does, you will be well prepared for it! It is quite a difficult question to respond but do use the following pointers when constructing your response.

- Get on same level as the child. This could be achieved by crouching or sitting down.
- Change vocabulary to suit the age of the child. This is of particular importance as it will make the child more responsive. If you talk in an authoritarian tone then this will probably make the child go into his/her shell.
- Consider using the child's favourite toy to make them feel more secure.
- Consider using the parents/guardian as reassurance and a communication aid.
- Start off by getting the child to explain the problem. Do not always act on the parent's assumptions, as these could be wrong. As a firefighter you will often arrive at incidents before the Ambulance Service arrives and will therefore have to carry out initial medical care and assessment.
- If child does not speak to you, consider asking the parents to get the answers to the questions for you.
- If appropriate, explain the actions that you are taking in order to reassure the child. For example, if placing a neck support on the child during a road traffic collision.
- Possibly take off your formal uniform jacket to make the child feel more at ease.
- Calm the parents down; this will calm the overall situation down and help them to assist you in dealing with the child.

SAMPLE GENERIC INTERVIEW QUESTION - NUMBER 20

How would you control an aggressive and confrontational person at an incident?

TIP: As a Firefighter you will need to deal with confrontational and aggressive people on a number of occasions. At no time should you ever put yourself in danger. The safety of yourself, your team and your equipment must always come first. You need to bear in mind that, on occasions when you are dealing people under the influence of alcohol, the Police may be a considerable time before attending. If you feel that any situation is become untenable, or that you are being placed in unnecessary danger, then you must always retreat. As a Firefighter you will learn how to carry out what is called a 'Dynamic Risk Assessment'. This will help you to determine how dangerous the situation is and what action you should take. Take a look at the following sample response.

"Whilst working as a door supervisor in a city centre bar, I was made aware by a customer that there was an altercation between a member of bar staff and a customer. As I went to the bar, I could see that there was a person shouting and swearing at the bar staff and manager.

I walked up and asked the aggressor in a clam manner what the problem was. His reply was "Go away, this has got nothing to do with you". My reply to the man was that I was the door supervisor and that it was something that I had to deal with. His aggression then turned to me and he began to shout and swear at me. I maintained a clam attitude but remained assertive and used open hand hands whilst dealing with the situation. I didn't block the exit as he could have seen this as me trying to intimidate him and it could therefore cause a fight or flight situation. I said that I had been fair with him and wanted to talk outside to find out his version of what had gone on. When we were outside, away from the noise and his friends and the bar staff, he stated that he had given the bar man a £20 note but only given change from £10. I told him to wait outside and I would go and see the bar staff and try and sort the matter out.

After I had spoken to the bar man and the manager, they agreed that he did in fact give him a £20 note and was then given the correct change. I asked the bar man to come outside and speak to the man in order to explain what had happened. They both shook hands and both carried on with their nights. I feel that I dealt with the situation effectively. I could have gone into the situation with an aggressive attitude, which would have probably escalated the situation and just made matters worse. However, through effective conflict management, I was able to successfully defuse the situation."

SAMPLE GENERIC INTERVIEW QUESTION - NUMBER 21

What do you think will be the best parts of the job?

This question is designed to assess your understanding of the role and it also gives the interviewer some idea of your intentions for joining. If you tell the panel that the best parts of the job are "Speeding down the road in a Fire Engine with lights and sirens going" or "Dealing with injuries and fatalities" then you are going to score very low for this question. Take a look at the following bullet points which will give you some great pointers for responding to this question.

- Helping the local community and educating them in resuscitation and first aid.
- Improving people's lives, well-being and in extreme situations, saving lives.
- Being a role model for the Fire Service and serving the community.
- Working with a professional and diverse group of people.
- Making a difference to the community in which I am working within.
- Learning new skills.
- Continually improving and developing as a person.
- The shift work. Not having to work in a regular 9-5 job can be very appealing to some people. (Please see the next question for how shift work can be negative.)
- Looking after your equipment and making sure that it is ready for operational use.
- Learning procedures and continually training to be prepared for every eventuality.
- Working with the other emergency services.
- Providing reassurance to the community not only physically but also mentally. The people we serve will feel safe and comforted knowing we are here when they need us.

SAMPLE GENERIC INTERVIEW QUESTION - NUMBER 22

What do you think will be the bad points about the job?

You obviously need to be careful when responding to this question, simply because there are some aspects to this role that cannot be avoided. For example, if you state that "seeing blood or fatalities" makes you feel uneasy, then you are not going to make it through the interview. Be careful when responding to the interview question and use the following pointers as a guide to assisting you in your preparation.

- The shift patterns will have a big impact on your life and your family's life and could be both mentally and physically demanding at times. Therefore you will require a level of resilience.
- Being in a position where you are not able to help someone or prevent the situation worsening.
- Not getting to people in time due to traffic congestion and people failing to move out of the way when you are responding to an incident under blue light conditions.
- Cars that are parked illegally that could hinder an emergency vehicle getting through.
- Hoax callers that waste time and also cost lives.

SAMPLE GENERIC INTERVIEW QUESTION - NUMBER 23

Why does the Fire Service have a uniform?

Whilst this is an unlikely question it is still worth reading the following advice.

TIP: There are numerous benefits to wearing a uniform. Not only does it allow you to be easily identifiable to members of the public and other emergency crews, but is also affords you a level of protection from injury during incidents. Take a look at the following pointers which you should try to include in any response that you create.

- A uniform is a form of Personal Protective Equipment (PPE). Your helmet will protect you from falls and overhead objects. Your tunic will protect you from dirt, splashes and spillages. Your steel toe capped shoes/boots will protect your feet and your gloves will protect your hands from spillages and waste etc.
- A uniform is a sign of strength, authority and it gives reassurance to the public.
- A uniform brings the service together as one large team, regardless or beliefs and differences. This also demonstrates the equality and diversity of the Fire and Rescue Service.
- The public also associate the red Fire Engine, yellow helmet and navy blue uniform with the Fire Service. This makes the them easily identifiable to the public, other emergency services and would also be a great benefit at a major incident. This principle is also adopted by other emergency services such as the Ambulance Service (green), Police (Blue/Black) and with HEMS (orange flight suits.)
- The helmet will protect you from falls and also overhead objects.
- Uniforms can also identify those people whom have senior rank. This is of particular importance at major incidents.
- A uniform is a sign of discipline within an organisation. It also requires pride to wear it which is a positive aspect.

SAMPLE GENERIC INTERVIEW QUESTION - NUMBER 24

What would you do if someone asked you to take your shoes off in their house?

Diversity is commonplace in today's society. I can remember serving as a Watch Commander at Gravesend Fire Station in Kent where the community was fantastically diverse. On many occasions we were asked to remove our footwear when entering people's homes that were from different backgrounds. Where possible, and in a non-fire situation, I respected the home owner's and asked my crew to remove their footwear. However, with incidents that involved fire or the need to operate equipment in the home I explained that this was not possible. I would then limit the number of persons who entered the home. Take a look at the following tips which will help you to understand how to react to this type of request.

- Be sympathetic to them for the reasons for asking to take them off (this could be for religious reasons or simply because they don't want their carpets becoming dirty). If it is for religious reasons then I would carry out a risk assessment and decide whether it was possible or not. I would always follow my training and procedural guidance.
- I would explain that I have to wear my shoes/boots for health and safety reasons.
- I would also explain that if I did take my shoes off this would delay me in reaching the incident/fire and this would affect me being effective, safe and prompt.

SAMPLE GENERIC INTERVIEW QUESTION - NUMBER 25

Why do we want the Fire Service to be diverse?

Any public service must be diverse in order to represent the community in which it serves. Society is diverse in nature and therefore it is only right that all public services are representative of the community. For example, there are many groups of people in our society that do not speak English. Therefore, it is only right that the service employs staff who are capable of speaking certain languages and also understanding different groups needs and requirements. Take a look at the following sample response.

"Society is a very diverse place. It has people from all over the world of different ages and capabilities from the poor to the rich and different religious beliefs and sexual preference. The Fire & Rescue Service must reflect this by employing people from diverse backgrounds. It also sets an example that other people and businesses will hopefully follow. This also shows leadership from the Fire Service. I feel it is important to show the public that you are diverse as an organisation. If the Fire Service were not diverse, maybe people from different cultures or that have less capabilities would not ring for help when needed and this goes against the services vision and values."

SAMPLE GENERIC INTERVIEW QUESTION - NUMBER 26

Can you provide an example of a situation when you have had to work under pressure?

The role of a Firefighter will often requirement to work under extreme pressure. Therefore, the recruitment staff will want to know that you have the ability to perform in such an environment. If you already have experience of working under pressure then you are far more likely to succeed and be capable of meeting the demands of the job. When responding to a question of this nature, try to provide an actual example of where you have achieved a task whilst being under pressure. Don't forget to follow the guidance at the beginning of this guide which related to responding effectively to 'situational' interview questions. Questions of this nature are sometimes included in the application form, so try to use a different example for the interview, if the question comes up.

Sample response

"Yes, I can. In my current job as car mechanic for a well-known company, I was presented with a difficult and pressurised situation. A member of the team had made a mistake and had fitted a number of wrong components to a car. The car in question was due to be picked up at 2pm and the customer had stated how important it was that his car was ready on time because he had an important meeting to attend. We only had two hours in which to resolve the issue and I volunteered to be the one who would carry out the work on the car. The problem was that we had 3 other customers in the workshop waiting for their cars too, so I was the only person who could be spared at that particular time. I worked solidly for the next 2 hours, making sure that I meticulously carried out each task in line with our operating procedures. Even though I didn't finish the car until 2.10pm, I managed to achieve a very difficult task under pressurised conditions whilst following strict procedures and regulations. I understand that the role of a paramedic will require me to work under extreme pressure at times and I believe I have the experience to achieve this. I am very meticulous in my work and always ensure that I take personal responsibility to keep up to date with procedures and policies in my current job."

SAMPLE GENERIC INTERVIEW QUESTION - NUMBER 27

What skills do you possess that you think would be an asset to our organisation?

When responding to questions of this nature try to match your skills with the skills that are required of a Firefighter. On some Fire Service websites you will be able to read about the type of person they are looking to employ, usually in the recruitment section. An example of this would be:

> If you're looking for a job with variety, a challenge and good career progression, why not think about becoming a firefighter? The alarm sounds. You pull on your firegear, get on the fire engine and race out double fast. This is the part everyone knows. But there's a lot more to being a firefighter. One of your most important roles will be to tell people about fire safety and fire prevention, particularly older people, children and people whose first language isn't English. You will also visit people in their homes to carry out fire safety visits, advising businesses and finding innovative ways to make sure our key safety messages get through. You'll also be expected to know the risks and potential hazards to be found in the area around your fire station. Constant practice, training in new skills, exercises and going to lectures are all essential to keeping your skills sharp as well as maintaining your physical fitness. And, of course, you need to deal calmly and professionally with emergencies like fires, road traffic collisions, chemical spills and floods, working alongside other emergency services and your colleagues to help people who are in distress.

Just by looking at the Fire Service's website you should be able to obtain some clues as to the type of person they are seeking to employ. It is also worthwhile studying the job description and person specification, as these will also provide details of the type of person they are looking to employ. Try to think of the skills that are required to perform the role you are applying for and include them in your response.

Sample response

"I am a very conscientious person who takes the time to learn and develop new skills correctly. I have vast experience working in a customer-focused environment and fully understand that excellent patient care is important. It is important that every member of the team works towards providing a high level of service. I believe I have the skills, knowledge and experience to do this. I am a very good team player and can always be relied upon to carry out my role to the highest of standards. I am a flexible person and understand that there is a need to be available at short notice to cover duties if required. In addition to these skills and attributes, I am a very good communicator and understand that different members of the community will need a different approach.

For example, when dealing with elderly members of the community I will have to be very patient and cater for their needs in a more sensitive manner. I am highly safety conscious and have a health and safety qualification to my name. Therefore, I can be relied upon to perform all procedures correctly and in line with my training and will not put others or myself in any danger whatsoever.

Finally, I am very good at learning new skills, which means that I will work hard to pass all of my continuation training if I am successful in becoming a paramedic."

SAMPLE GENERIC INTERVIEW QUESTION - NUMBER 28

Can you provide us with an example of a safety-related task that you have had to perform?

Safety is an extremely important part of the firefighters role and the recruitment staff need to know that you are capable of working safely at all times. The term 'safety' should be an integral part of your responses during the interview. Making reference to the fact that you are aware of the importance of safety is a positive thing.

When responding to safety-related questions you should try to include examples where you have had to work to, or follow, safety guidelines or procedures. If you have a safety qualification then it is definitely worthwhile mentioning this during your interview. Any relevant safety experience or safety-related role should also be discussed.

Sample response

"I currently work as a gas fitter and I am often required to perform safety-related tasks. An example of one of these tasks would involve the installation of gas-fired boilers. When fitting a gas boiler, I have to ensure that I carry out a number of safety checks during the installation stage, which ensures my work is safe and to a high standard. In addition to carrying out work in line with procedures and regulations, I also carry out daily checks on my equipment to ensure that it is serviceable, operational and safe. If I find any problems then I immediately take steps to get the equipment repaired by a qualified engineer or technician. I have been trained, and I am qualified, to carry out my work in accordance with strict safety guidelines. I also have a number of safety certificates to demonstrate my competence. I am fully aware that if I do not carry out my job in accordance with safety guidelines there is the possibility that somebody may be injured or even killed."

SAMPLE GENERIC INTERVIEW QUESTION NUMBER 29

Do you think you would get bored of routine tasks such as checking your equipment and reading up on procedures?

Of course the only answer here is no. Part of the job of a firefighter is to check and familiarise yourself with your equipment and keep up to date with procedures. Every job has mundane tasks but it is usually these tasks that are the most important.

Sample response

"Absolutely not. I would take just as much care and attention mopping the fire station floor as I would attending operational incidents. I understand that there will be routine tasks to carry out every day and I would always make sure that I carry them out with pride and professionalism. I already have experience of routine tasks in my current job as a car mechanic. Everyday I have to carry out an inventory and serviceability assessment of my tools. Whilst the task is mundane I always ensure that it is done correctly as someone's life could depend upon it."

SAMPLE GENERIC INTERVIEW QUESTION NUMBER 30

How many people work for this organisation and where are the fire stations located?

Questions that relate to facts and figures or the structure of the service are commonplace. The panel will want to know that you are serious about joining their Service and that you have looked into their organisation in detail. Make sure you study the organisation, the people and its structure before you attend the interview. You will be able to find plenty of information on the service's website. You may also decide to visit your local fire station to obtain some facts and figures from the crews.

SAMPLE GENERIC INTERVIEW QUESTION NUMBER 31

Have you ever worked during the night and how do you feel about working shifts?

The work of a firefighter involves irregular shift patterns and the service will want to know that you can handle them. Speak to any person who works shifts and they will tell you that after a number of years they can start to take their toll. Remember to tell the panel that you are looking forward to working shifts and, in particular, night duties. If you can provide examples of where you have worked irregular shift patterns then remember to tell them as this will work in your favour. It may also be advisable to tell the panel that your family fully support you in your application and they appreciate the impact working shifts may have on your home and social life.

SAMPLE GENERIC INTERVIEW QUESTION NUMBER 32

Can you provide us with an example of a project you have had to complete and the obstacles you had to overcome?

Having the ability to complete tasks and projects successfully demonstrates that you have the ability to complete your paramedic/ ambulance technician training.

Many people give up on things in life and they fail to achieve their goals. The recruitment staff will want to know that you are going to complete all of your training successfully and, if you can provide evidence of where you have already done this, then this will go in your favour.

When responding to this type of question, try to think of a difficult, long drawn-out task that you achieved despite a number of obstacles that were in your way.

CHAPTER 3
FINAL INTERVIEW TIPS & ADVICE

Within this final section of the guide I will provide you with some useful tips that will help you prepare fully for the interview.

PRE-INTERVIEW PREPARATION AND HOW I WOULD PREPARE FOR THE FIREFIGHTER INTERVIEW

During my pre-interview preparation I will concentrate on developing my interview technique. This will involve concentrating on the following key areas:

- Creating a positive first impression
- Presentation
- Effective communication
- Body language and posture
- Final questions
- Creating a positive final impression

Let's now break down each of these areas and look at them in detail.

Creating a positive first impression

An interview panel will naturally create a first impression of you. As soon as you walk into the interview room they will be forming an opinion. Therefore, it is important that you get off on the right foot. Whenever I walk into any interview room I will always follow this process:

Knock before I enter the room

Walk into the interview room standing tall and smiling

Stand by the interview chair and say
"Hello, I'm Richard, pleased to meet you."

Shake the hand of each interviewer firmly, whilst looking them in the eye

Sit down in the interview chair, only when invited to do so

Sit in the interview chair with an upright posture and with my hands resting palms facing downwards on the top of my knees, feet firmly on the floor

By following the above process I will be creating a positive first impression and demonstrating good qualities such as manners, self-discipline, politeness and motivation. Many people ask me whether it is good practice to shake the hands of the interviewer. My advice is this: I would recommend shaking the hands of each interviewer at the end of the interview as opposed to the beginning. Make sure you give a firm handshake and look the person in the eyes.

Presentation

Presentation effectively means how I intend to dress for the interview and also how I intend to come across. I want the interview panel to see me as a professional, motivated, conscientious and caring person who is taking the interview very seriously.

Some Fire and Rescue Services do not require you to dress formally for the interview. For some bizarre reason some senior managers believe that a person should not be assessed on how they present themselves at interview. Personally, I disagree with this approach entirely. Whilst I agree there is no need to go out and buy an expensive suit or new pair of shoes, I do believe that a potential employee should make an effort in their appearance. I therefore recommend that you wear a formal, clean and well pressed suit with clean and polished shoes.

For the interview I will make sure that my suit is cleaned and pressed, my shoes are polished, and my personal hygiene is up to standard. This means simple things such as taking a shower, shaving, having a haircut and general grooming. I will always avoid brightly coloured clothes and generally go for a conservative approach such a dark blue, black or grey suit. If I do decide to wear any brighter, more vibrant colours, then this will be in form of a tie. I would strongly advise that you avoid brightly coloured socks or ties with cartoon characters on them!

A good applicant

A good applicant is someone who has taken the time to prepare. They have researched both the organisation they are applying to join and also the role that they are being interviewed for. They may not know every detail about the organisation and the role but it will be clear that they have made an effort to find out important facts and information. They will be well presented at the interview and they will be confident, but not overconfident. As soon as they walk into the interview room they will be polite and courteous and they will sit down in the interview chair only when invited to do so. Throughout the interview they will sit upright in the chair and communicate in a positive manner. If they do not know the answer to a question they will say so and they won't try to waffle. At the end of the interview they will ask positive questions about the job or the organisation before shaking hands and leaving.

A poor applicant

A poor applicant could be any combination of the following. They will be late for the interview or even forget to turn up at all. They will have made little effort to dress smartly and they will have carried out little or no preparation. When asked questions about the role they will have little or no knowledge. Throughout the interview they will appear to be unenthusiastic about the whole process and will look as if they want the interview to be over as soon as possible. Whilst sat in the interview chair they will slouch and fidget. At the end of the interview they will try to ask clever questions that are intended to impress the panel.

Improving interview technique

How you present yourself during the interview is important. Whilst assessing candidates for interviews I will not only assess their responses to the interview questions but I will also pay attention to the way they present themselves. A candidate could give excellent responses to the interview questions but if they present themselves in a negative manner then this can lose them marks.

Take a look at the following diagrams, which indicate both poor technique and good technique.

Poor interview technique

The candidate appears to be too relaxed and casual for an interview.

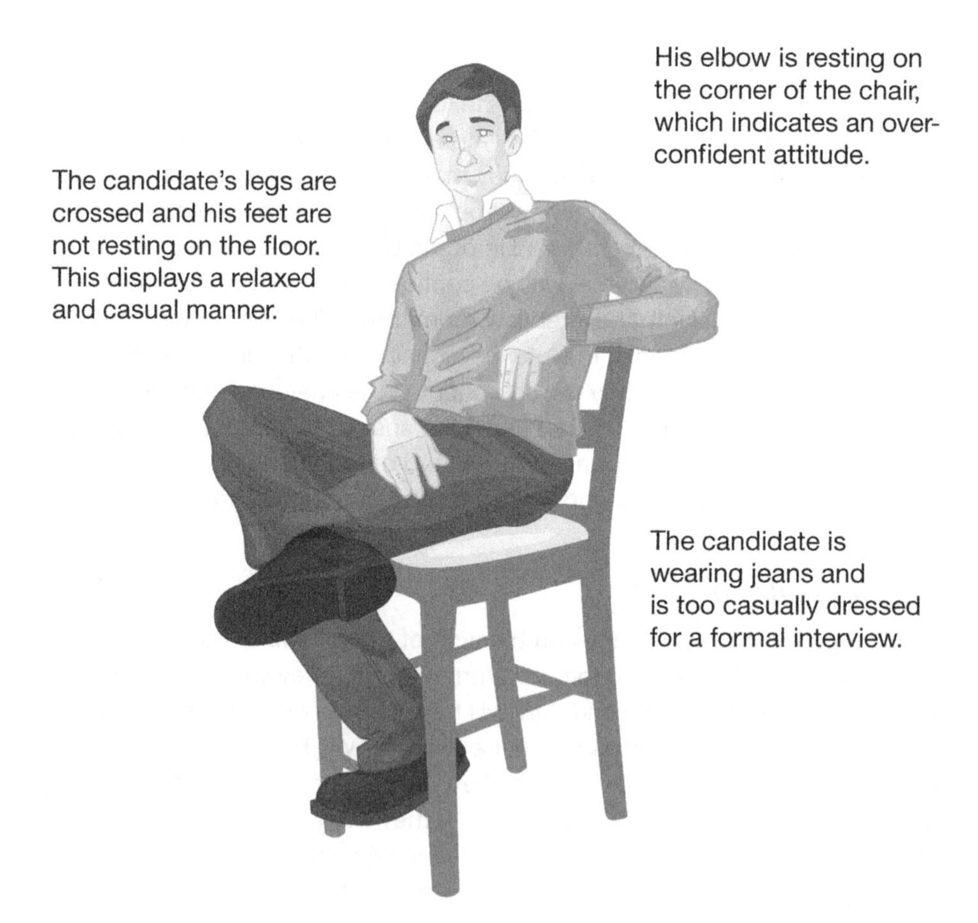

Good interview technique

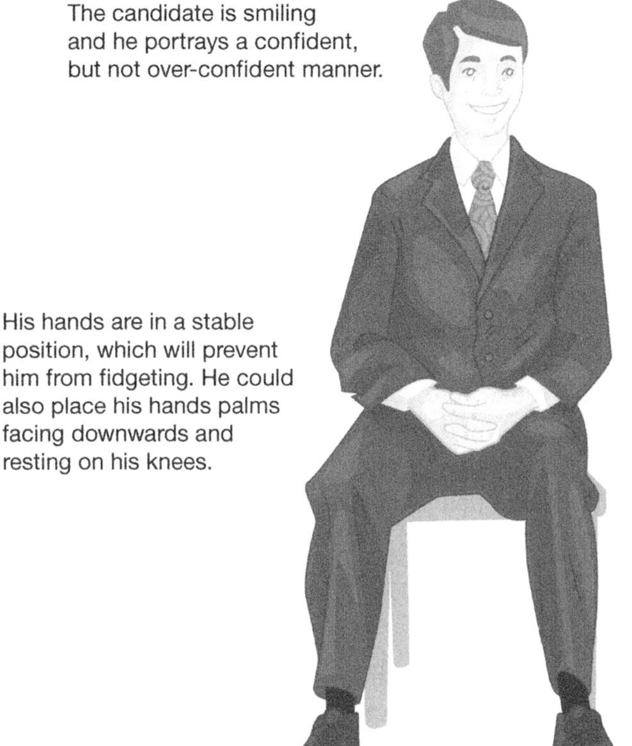

In the build up to your interview practise a few mock interviews. Look to improve your interview technique as well as working on your responses to the interview questions.

Effective communication

During the firefighter interview you will be assessed in how effectively you communicate. Effective communication is all about how you speak to the interview panel, how you structure your responses and also how you listen to what they have to say. When responding to the interview questions you should speak clearly and concisely, avoiding all forms of waffle, slang or hesitations such as 'erm'. Look at each interview panel member when answering each question. Even though an interview question will be asked by one member of the panel at a time, you should always respond to the entire panel collectively. Look them in the eyes when speaking to them but never stare them out. This will only portray you in an aggressive or confrontational manner.

If you are unsure about a response to an interview question then just be honest. Consider saying something along the lines of:

"I'm sorry I do not know the answer to that question. I will look the answer up as soon as I get back home and contact you to let you know the answer."

If they accept this response, make sure you do research the response and contact them to let them know. When the interview panel are speaking to me, or if they are asking me a question, I will always demonstrate good listening skills. This means that I will use facial expressions to show that I am taking on-board what they are saying and I will also nod to show them that I understand the question(s).

Body language and posture

Whilst sat in the interview I will always make a conscious effort to sit upright and not slouch in the chair. I personally like to use my hands to emphasise points when responding to the questions but I will be careful not to overdo it. Even if the interview is going great and you are building up a good rapport with the panel, don't let your standards drop. Always maintain good body language and posture for the duration of the interview.

Final questions

Before I attend the interview I will always think of two questions to ask the panel at the end. However, don't be trapped in the thinking that you must ask questions. It is acceptable to say:

"Thank you but I don't have any questions. I have already carried out lots of research and you have answered some of my questions during the interview."

Some people believe that you must ask three, four or even five questions at the end of the interview – this is total nonsense. Remember that the interview panel will have other people to interview and they will also need time to discuss your performance.

If you do decide to ask questions then make sure they are relevant.

Here are a couple of good questions to ask at the end:

"Is there any literature or information I can read about the Fire Service whilst waiting to find out the results of my interview? I would like to start preparing for the trainee course just in case I am successful."

"Whilst researching your website I noticed that you recently carried out a campaign on cooking safely in the home. I was wondering how successful the campaigns are and whether firefighters are allowed to think of ways for helping the organisation to come up with ways to reduce fires and deaths in the home?"

Creating a positive final impression

I believe that a final positive statement at the end of the interview can work wonders:

"I just want to say thank you for inviting me along to interview. I've really enjoyed the experience and I have learnt a tremendous amount about your Fire Service. If I am successful then I promise you that I will work very hard in the role and I will do all that I can to surpass your expectations."

FINAL THOUGHTS

- If you have prepared yourself fully leading up to the interview you will hopefully have the confidence to perform to the best of your ability on the day. Preparation is key to your success so take the time to follow the instructions and guidance provided within this section.
- Make sure you know the correct date, time and location of your interview and be there early, with plenty of time to spare.
- Take into account the possibility of heavy traffic, a breakdown and parking etc.
- It is a good idea to make sure you know exactly where you are going. I recommend you visit the interview location prior to the day so you are familiar with the location and how to get there.
- Ensure you have revisited your application form. They may ask you questions about its content so make sure you know what you submitted.
- Ensure you know both about the role of the firefighter and information relating to the actual Fire and Rescue Service you are applying to join.
- Have a good understanding about Community Fire Safety before attending the interview.
- Take a look at their website and find out what is current, such as Community Fire Safety campaigns, Integrated Risk Management etc.
- It is always a good idea to arrange a visit to a fire station before your interview. Ask them questions about their role and their working day so that you are fully prepared for your interview.
- Have knowledge of Health and Safety and in particular the 5 steps to Risk Assessment.
- Be aware of the Race Equality Scheme for the Fire Service you are applying to join. You could be asked a question about equality and fairness during your interview.
- Be aware of the Personal Qualities and Attributes that relate to the role of the firefighter. These are the areas that you will be assessed against. You must be able to provide specific examples of each key area.

- Make sure you dress smartly. Image is important in any interview and demonstrates that you are serious about the whole process. If you turn up in jeans and trainers the interviewer may view this as negative. If you want more interview advice then you can purchase the 'How to pass the firefighter interview' DVD through the website www.how2become.co.uk.
- Check if you are required to take anything with you such as references, certificates, driving licence or proof of your educational qualifications.
- Remember to smile during the interview and be positive.
- Think of two possible questions to ask at the end of the interview. Try not to be clever when asking questions but instead ask ones that are relevant such as "Where is the organisation planning to go with Community Fire Safety".
- When you enter the interview room make sure you are polite and say hello, good morning or good afternoon. Saying nothing at all will come across as being rude.
- Don't sit down until invited to do so. Whilst this is not essential it does demonstrate good manners.
- Make sure you sit comfortably and don't slouch. A good posture will speak volumes about your confidence and determination to succeed.
- Think before you answer any questions. There is nothing wrong with pausing for a second to think about your answer. If you are unsure, ask them to repeat the question.
- Look interested when they are asking you questions and be positive in your answers.
- If you are unsure of an answer try not to 'waffle' or make something up. If you can't answer a question just be honest and move on.
- Speak up when answering any questions and make positive eye contact. This doesn't mean staring out the interviewer!
- Don't over use your hands. Some hand movement or expression is good but too much can be distracting.

- The interviewers may ask you more generic questions relating to your past experiences or skills. These may be in relation to how you solve problems, your strengths and weaknesses, team-working skills, communication skills and questions that relate to the physical aspects of the role. Make sure you have examples for each of these.
- Try to speak to a current serving firefighter of the service that you are applying to join. Ask him/her what it is like to work for that particular service and what current issues they are facing.
- Try to think of a time when you have made a mistake and how you learnt from the experience. The panel may ask you questions that relate to how you deal with setbacks in your life.
- When you complete the application form, make sure you keep a copy of it. Before you go to your interview ensure that you read the application form over and over again as you may find you are asked questions about your responses.
- Don't be afraid to ask the interviewer to repeat a question if you do not hear it the first time. Take your time when answering and be measured in your responses. You will be assessed against effective communication so try out your responses in a mock interview setting prior to your big day.
- If you don't know the answer to a question then be honest and just say 'I don't know'. This is far better than trying to answer a question that you have no knowledge about. Conversely, if your answer to a question is challenged, there is nothing wrong with sticking to your point but make sure you acknowledge the interviewer's thoughts or views. Be polite and never get into a debate.

how2become